RISING WAVES

Unveiling Japan's Quake Chronicles

A Data-Driven Journey into Earth's
Tremors, Tragedy, and Triumphs

DE GIST LOVERS

Table of contents

Introduction

In the early hours of the first day of the year 2024, Japan found itself thrust into the throes of a seismic ordeal. A series of powerful earthquakes, originating in the Sea of Japan off the coast of Ishikawa and neighboring prefectures, shook the nation to its core. The tremors, reaching a peak magnitude of 7.6, were a stark reminder of the geographical vulnerability that Japan has historically grappled with.

This seismic saga unfolded shortly after 4 pm local time (7 am UK time), catching both residents and authorities off guard. The Ishikawa prefecture, situated along the center of Japan's western coast, became the epicenter of this unnerving event. With its 581-kilometer coastline bordered by the Sea of Japan, Ishikawa is no stranger to the tremors that periodically visit the region. The latest episode, however, proved to be a formidable challenge, leaving the prefecture grappling with the aftermath.

As the nation rang in the new year, thousands of residents were abruptly thrust into a state of urgency. Tsunami warnings initially sent shockwaves through coastal communities, prompting swift evacuation efforts. The entire nation stood at the precipice of a potential disaster, anxiously awaiting updates from the Japan Meteorological Agency.

In the wake of this unforeseen catastrophe, the unfolding narrative painted a harrowing picture — collapsing buildings, a rising death toll, and a landscape scarred by the relentless force of nature. Fires ignited in the aftermath, and landslides further compounded the challenges faced by both residents and emergency responders.

The earthquakes' impact was not confined to physical structures; it reverberated through the collective psyche of the nation. The fragility of life, the resilience of communities, and the indomitable spirit of the Japanese people took center stage in the unfolding drama. As the world watched, Japan faced a daunting task — to navigate through the chaos, mourn the lives lost, and rebuild amidst the aftershocks that lingered in the air.

The significance of seismic activities in the region

Nestled within the Pacific Ring of Fire, Japan stands as a testament to the raw power of seismic activities. This horseshoe-shaped zone, characterized by high seismic and volcanic activity, encompasses the Pacific Ocean and harbors some of the most geologically dynamic areas on Earth. For Japan, situated at the convergence of four major tectonic plates – the Pacific, Philippine Sea, Eurasian, and North American plates – seismic activities are not mere geological events but rather an integral part of its complex landscape.

The significance of seismic activities in the region stems from the intricate dance of these tectonic plates beneath the Earth's surface. The constant interaction and collision of these plates create immense pressure, leading to the release of energy in the form of earthquakes. This geological phenomenon, while posing a constant threat, has also shaped Japan's topography and cultural identity over centuries.

In Japan, earthquakes are not just occasional disruptions; they are a part of the national consciousness. The historical context is rich with seismic events that have left an indelible mark on the nation. From the devastating Great Kanto Earthquake of 1923 to the more recent horrors of the 2011 Tohoku earthquake and tsunami, Japan has faced the relentless challenge of reconciling its existence with the unpredictable forces beneath its feet.

The significance extends beyond geological fascination; it permeates every aspect of life. The architectural resilience of Japanese buildings, stringent building codes, and an unparalleled early warning system are all reflections of a society deeply attuned to the seismic rhythm of its surroundings. The seismic activities, while posing threats, have also catalyzed advancements in earthquake preparedness, making Japan a global leader in disaster resilience.

As "Rising Waves" ventures into the heart of Japan's seismic narrative, it unravels not just the geological implications but the profound cultural and societal resonance of living in a region where the ground beneath is in perpetual motion. This significance, ingrained in the collective memory of the nation, adds layers of complexity to the unfolding drama of the recent earthquakes in Ishikawa. The earthquakes are not isolated events but threads woven into the larger tapestry of Japan's seismic saga, a saga that continues to shape its destiny.

In the aftermath of the recent earthquakes that shook Japan to its core, a unique opportunity arises — a chance to delve into the seismic events with a data-driven lens. "Rising Waves" sets the stage for an exploration that transcends the narratives of tragedy and resilience, venturing into the realm of analytics and insights.

As the nation grapples with the immediate aftermath, the data becomes a beacon of understanding in the chaos. Beyond the headlines and emotional accounts, lies a wealth of information waiting to be deciphered. The seismic activity, captured in numerical magnitudes, geographical coordinates, and historical patterns, tells a story of its own. This data-driven exploration seeks to unravel the mysteries concealed within the trembling ground.

The seismic data, often relegated to scientific reports, takes center stage in "Rising Waves." It becomes a protagonist in its own right, guiding readers through the intricacies of plate tectonics, fault lines, and the geological nuances that set the stage for the recent earthquakes. Through this exploration, readers are invited to witness the seismic puzzle, to understand the patterns that precede such events, and to appreciate the intricate ballet of forces beneath the Earth's surface.

But the data doesn't stop at the geological; it extends to human responses, emergency preparedness, and the efficacy of warning systems. Analyzing the evacuation timelines, the impact on different regions, and the effectiveness of response strategies becomes a crucial part of the narrative. As the story unfolds, readers are presented with a comprehensive view — a synthesis of human experience and numerical precision.

"Rising Waves" stands at the intersection of storytelling and analytics, where each piece of data contributes to a deeper understanding of the seismic events. The stage is set not just for a recounting of events but for an exploration that enlightens, educates, and empowers. It is a journey guided by data, where the numbers become storytellers, and the analytics become a roadmap through the seismic labyrinth that Japan navigates with each passing tremor.

Chapter 1:

Prelude to Disaster

Historical context of earthquakes in Japan

Japan's history is etched with the seismic imprints of its volatile geological setting. To comprehend the recent earthquakes, one must delve into the historical context, where the echoes of past tremors reverberate through time, shaping the nation's resilience and preparedness.

The seismic saga in Japan finds its roots in the devastating Great Kanto Earthquake of 1923. With a magnitude of 7.9, it left Tokyo and surrounding areas in ruins. The toll was not just in lives lost but in the profound impact on urban planning, architecture, and disaster response.

The city of Kobe became synonymous with resilience following the 1995 earthquake. At magnitude 6.9, it unleashed destruction, claiming over 6,000 lives. The aftermath, however, prompted a paradigm shift in earthquake preparedness, influencing building codes and emergency response nationwide.

The Tohoku earthquake and tsunami of 2011, with a staggering magnitude of 9.0, stands as one of the most powerful ever recorded. The devastation underscored the vulnerability of coastal regions but also showcased the effectiveness of Japan's early warning systems and community preparedness.

Japan's location within the Pacific Ring of Fire amplifies its seismic vulnerability. The constant interactions of tectonic plates generate a symphony of seismic activity, where the undulating dance beneath the Earth's crust creates both beauty and chaos.

The Pacific Plate's subduction beneath the North American, Philippine Sea, and Eurasian Plates creates a complex web of fault lines. The seismic ballet involves frequent tremors as these plates grapple for space, setting the stage for both moderate quakes and catastrophic events.

The historical seismic scars prompted Japan to pioneer earthquake-resistant architecture. From flexible foundations to shock absorbers, the nation's buildings are a testament to engineering ingenuity, continuously evolving to withstand the relentless shaking.

Beyond structures, Japan has invested in community preparedness. Regular drills, education on evacuation procedures, and a culture that values collective safety contribute to a society ready to face the seismic challenges thrown its way.

Earthquakes have permeated Japan's cultural psyche. From folklore to art, seismic events find expression in myriad forms. The awareness of living on seismically active land is not just scientific but deeply woven into the fabric of daily life.

Japan's seismic history is a narrative still in the making. Each tremor, a page in a chronicle that testifies to the nation's endurance, adaptability, and an unwavering commitment to learning from the echoes of its seismic past.

"Rising Waves" unfolds against this backdrop, where the historical context becomes more than a mere prologue; it becomes the very foundation upon which Japan stands as it faces the seismic uncertainties of the present and the future.

The May 2023 earthquake in Ishikawa

Amidst the seismic tapestry of Japan's history, a poignant note was added in May 2023 when Ishikawa, the very region now grappling with the recent earthquakes, experienced its own tremor. This seismic prelude, while not as formidable as the recent events, foreshadowed the vulnerability of the area.

With a magnitude of 6.5, the May 2023 earthquake served as a gentle reminder of the restless forces beneath Ishikawa's surface. Though not causing widespread devastation, it claimed at least one life, underlining the region's susceptibility to seismic events.

In the aftermath of the May 2023 quake, Ishikawa took steps to fortify its resilience. The event prompted a reassessment of emergency response strategies, building structures, and community preparedness, as the region braced itself for the possibility of more intense seismic activity in the future.

The May 2023 earthquake became a valuable lesson in the ongoing seismic education of Ishikawa. Insights gained from this event contributed to the region's ability to respond more effectively when faced with the recent, more powerful earthquakes. The echoes of the past guided the present response.

While the May 2023 earthquake may have been a mere tremor in comparison to recent events, it cast a shadow of anticipation over Ishikawa. The region's experience with seismic activity became a crucial prologue to the unfolding narrative of "Rising Waves," as it grapples once again with the seismic uncertainties that define its existence.

Ishikawa's geography and its susceptibility to earthquakes

Nestled along Japan's western coast, Ishikawa unfolds as a region both blessed and burdened by its geographical setting. With a coastline stretching 581 kilometers along the Sea of Japan, its allure lies in the maritime beauty that defines its borders.

The Sea of Japan, while providing breathtaking vistas, also introduces a precarious element. Ishikawa finds itself in the midst of tectonic intricacies, where the Pacific, Philippine Sea, and Eurasian plates converge. This collision results in the creation of fault lines, setting the stage for seismic activities that punctuate the region's history.

Recent events echo the tectonic intricacies beneath Ishikawa's surface. The recent earthquakes, though jolting, are part of a seismic puzzle—an unpredictable manifestation of the delicate equilibrium maintained by the region.

In the towns and cities dotting Ishikawa's landscape, seismic uncertainty is woven into the fabric of daily life. From the urban expanse of Kanazawa to the coastal communities lining the Sea of Japan, every corner of Ishikawa bears witness to the dual narrative of beauty and vulnerability.

Yet, Ishikawa stands resilient. Generations molded by seismic coexistence have cultivated a spirit of preparedness. Architectural innovations, community drills, and a collective understanding of the seismic rhythm empower Ishikawa to face the uncertainties presented by the Earth.

Chapter 2:

The Unfolding Tragedy

Detailed account of the series of earthquakes

A series of major earthquakes originating in the Sea of Japan off the coast of Ishikawa, with the largest registering a magnitude of 7.6, reverberated through the region, setting the stage for a night of uncertainty and chaos.

The Japan Meteorological Agency reported the initial tremors shortly after 4 pm local time, sending shockwaves through Ishikawa and neighboring prefectures. The region, no stranger to seismic activity, was thrust into the spotlight once again. The Ishikawa prefecture, situated at the center of Japan's western coast, became the epicenter of an unfolding drama that would grip the nation.

Tsunami warnings flashed across screens, prompting immediate evacuations as coastal communities braced for the possibility of towering waves. The initial fear of a catastrophic tsunami gradually shifted to a more manageable advisory, but the lingering threat loomed large. As the night unfolded, a complex series of events painted a vivid picture of the seismic chaos that gripped Ishikawa.

The human toll became apparent with each passing update. Reports of collapsed buildings, tragic fatalities, and the valiant efforts of firefighters grappling with at least 30 collapsed structures underscored the severity of the situation. A man lost his life as a building crumbled in Shika Town, a stark reminder of the physical devastation wrought by the seismic forces.

The landscape bore the scars of the earthquakes. A highway in western Japan became impassable due to landslides and road collapses. Images captured from a plane revealed a raging fire in Ishikawa City, where smoke billowed from buildings engulfed in flames. The aftermath unveiled crushed cars, fallen signs, and flooded areas, painting a somber tableau of the region's struggles.

The seismic onslaught continued, with more than 100 earthquakes and aftershocks recorded in the past 12 hours, ranging from magnitudes 7.6 to 2.9. The majority clustered near Noto in Ishikawa, the epicenter of the strongest quake. However, the tremors were felt in different parts of Japan, emphasizing the widespread impact.

Evacuation efforts were underway, with over 97,000 people urged to seek higher ground in nine prefectures along the western coast. Residents, grappling with fear and uncertainty, sought refuge in government offices and evacuation centers, laying on floors and watching news coverage of the unfolding disaster.

As the night wore on, U.S. President Joe Biden expressed solidarity with Japan, offering assistance to the affected people. The unfolding events signaled not only a physical struggle against seismic forces but also a diplomatic acknowledgment of shared humanity in the face of adversity.

In the midst of this turmoil, the seismic puzzle expertly analyzed by earthquake engineer Professor Anastasios Sextos offered a glimmer of hope. The peculiar offshore location of the earthquake, he explained, potentially reduced the exposure of major cities to stronger shaking, mitigating the impact. He spoke of Japan's earthquake preparedness, highlighting the country's leading role and its excellent early warning system.

The night unfolded as a harrowing saga of tremors, fear, and resilience

The initial 7.6 magnitude quake off the coast of Ishikawa

The epicenter of this geological convulsion was pinpointed in the Sea of Japan, just off the serene coast of Ishikawa. With a magnitude of 7.6, this initial quake sent shockwaves that would reverberate through the night, setting the stage for a series of seismic events that would unfold in relentless succession.

The Japan Meteorological Agency, the vigilant guardian against the unpredictable forces of nature, swiftly reported the seismic disturbance shortly after 4 pm local time. The tremors, originating in the depths of the Sea of Japan, unleashed their energy with a potency that hadn't been witnessed in Ishikawa for over four decades.

The significance of this initial quake lay not just in its immediate impact but in the ominous foreshadowing of a night fraught with uncertainty. Tsunami warnings, akin to urgent calls from a sentinel, flashed across screens, urging residents to evacuate coastal areas in anticipation of possible colossal waves. The specter of the devastating 2011 Tohoku earthquake and tsunami lingered in the collective memory, adding an extra layer of dread to the unfolding drama.

The Ishikawa prefecture, accustomed to seismic ripples in its history, found itself at the epicenter of attention. The magnitude 7.6 quake, a powerful shudder beneath the

Earth's surface, was a stark reminder of the region's vulnerability to the capricious dance of tectonic plates.

Impact on buildings, infrastructure, and daily life

The seismic onslaught that unfolded off the coast of Ishikawa on the first day of 2024 left an indelible mark on buildings, infrastructure, and the daily lives of those residing in the affected regions. As "Rising Waves" delves into the aftermath, a tapestry of destruction and resilience emerges, painting a vivid picture of the seismic impact.

The force of the earthquakes wreaked havoc on structures, leaving a trail of collapsed buildings in its wake. Reports from the crisis management team in Ishikawa revealed the grim reality—buildings reduced to rubble, their structural integrity compromised by the relentless shaking. Tragically, lives were lost as a man succumbed to the collapse of a building in Shika Town. The images captured in the aftermath showcased the fragility of the built environment in the face of nature's fury.

Roads, highways, and other vital infrastructures bore the brunt of the seismic onslaught. A highway in western Japan became impassable due to landslides and road collapses, disrupting the normal flow of transportation. The Noto Airport, a crucial lifeline for the region, canceled all flights, underscoring the immediate impact on essential services. The cracks in airport runways and the cancellation of air travel further underscored the vulnerability of critical infrastructure.

For the residents of Ishikawa and neighboring prefectures, daily life transformed into a surreal dance between fear and resilience. Evacuation orders disrupted routines as over 97,000 people were urged to seek refuge in higher ground. Crowded evacuation centers and government offices became makeshift shelters, with people lying on floors, anxiously watching news coverage of the unfolding disaster. The quest for basic necessities transformed into a challenge, with crowds flocking to shops, creating a chaotic scene as they sought water, bread, and rice amidst the aftershocks.

The seismic impact extended beyond the physical realm, seeping into the emotional and psychological fabric of daily life. The constant threat of aftershocks and the looming uncertainty about the future cast a shadow over the resilience of communities. Yet, amidst the chaos, glimpses of solidarity emerged. The U.S. President, Joe Biden, extended a hand of support, emphasizing the deep bond of friendship between nations in times of crisis.

The tragic loss of lives

Amidst the seismic chaos that unfolded off the coast of Ishikawa, the human toll became a poignant focal point—a heartbreaking chapter in the unfolding narrative of "Rising Waves." Four reported casualties stand as tragic reminders of the fragility of life in the face of nature's relentless forces.

The crisis management team in Ishikawa confirmed the devastating news: four lives lost in the wake of the earthquakes. Among them was an elderly man in Shika Town, pronounced dead after a building collapsed—a stark illustration of the immediate and catastrophic consequences of the seismic tremors.

The magnitude 7.6 quake, with its epicenter in the Sea of Japan, sent shockwaves that reverberated through the region, claiming lives and leaving families shattered in its wake. The toll of the tragedy extended beyond mere statistics; it became a somber reflection of the human cost of living in a seismically active zone.

As "Rising Waves" pays tribute to the four reported casualties, it does so not just in recognition of their untimely demise but as a testament to the vulnerability of communities facing the unpredictable forces of nature. Each life lost becomes a poignant reminder of the profound impact of seismic events, urging readers to empathize with the grief and anguish experienced by those directly affected by the earthquakes. The tragic loss of lives, etched into the narrative, serves as a solemn backdrop against which the resilience and strength of the human spirit in the face of adversity can be fully appreciated.

Chapter 3:
The Human Side

Personal stories of those affected

As "Rising Waves" unfolds, it weaves a tapestry of personal stories—intimate narratives that humanize the seismic chaos that befell Ishikawa. These stories, drawn from the experiences of those directly affected by the earthquakes, offer a glimpse into the raw emotions, resilience, and shared humanity that emerged in the face of adversity.

Ayako Daikai's Evacuation

In the bustling primary school in Kanazawa, Ayako Daikai found herself surrounded by the unfamiliar—classrooms, stairwells, hallways, and the gym packed with people seeking refuge. Evacuating with her husband and children soon after the earthquake hit, Ayako, a mother-of-two, revealed the emotional turmoil of uncertainty. "I also experienced the Great Hanshin Earthquake, so I thought it would be safest to evacuate," she shared. The decision to leave home and the undetermined timeline for return became emblematic of the choices made by families navigating the seismic unpredictability.

The Shrine in Toyama City

In Toyama City, a 70-year-old man, Daniel Smith, was among a thousand people gathered at a shrine, partaking in a New Year's tradition to pray for good luck. The seismic symphony commenced slowly, then escalated into violent shaking. "At first, people just were stunned and they kept trying to go to the shrine," Daniel recounted. The abrupt shift from tradition to chaos encapsulated the disorienting experience faced by individuals caught in the midst of the seismic upheaval.

Eyewitness accounts of the tremors

In the midst of the seismic tumult that engulfed Ishikawa on that fateful day, eyewitness accounts emerged as poignant testimonials to the raw and unrelenting force of the tremors. "Rising Waves" captures these firsthand experiences, providing readers with a visceral connection to the immediacy and intensity of the seismic events.

Daniel Smith, a 70-year-old man, found himself amidst a thousand people at Hie Jinja Shrine in Toyama City. As the initial tremors commenced, he recalled the surreal unfolding of events. "The first tremor started very slowly and everybody kind of left it off... And then it's just a violent shake, I mean violent shaking," he recounted. The transition from anticipation to the sudden and intense shaking became a vivid snapshot of the disorienting nature of seismic upheaval. The urgency in his voice as he described the violence of the shaking added a personal and emotional dimension to the unfolding chaos.

A video captured the moment when the seismic waves reached a temple in Kanazawa. Dozens of people, gathered to pray for good luck as part of a New Year's tradition, became unwitting participants in the seismic drama. The visual documentation of the temple's response to the shaking—swaying and vibrating in defiance of its solid structure—offered a visceral portrayal of the seismic forces at play.

Rescue efforts and the challenges faced by emergency services

The response to the seismic events becomes a testament to the resilience and dedication of those on the front lines, navigating a landscape transformed by destruction.

Firefighters, clad in their protective gear, found themselves thrust into a maelstrom of collapsed buildings. Reports indicated that at least 30 structures succumbed to the seismic forces in Ishikawa. The challenges were immediate and dire—rescue efforts focused on locating and extricating individuals trapped beneath the rubble. The urgency of the situation was underscored by the tragic loss of life, with an elderly man pronounced dead after a building collapsed in Shika Town.

Aerial photos revealed the extent of a huge fire sparked by the earthquakes in Ishikawa City. The searing flames engulfed several buildings, creating a daunting challenge for firefighters attempting to contain the blaze. Smoke billowed from the affected area, adding a layer of complexity to the already arduous rescue and firefighting efforts.

The seismic upheaval triggered landslides and road collapses, rendering a part of National Route 249 impassable. Emergency services found themselves contending not only with collapsed structures but also navigating disrupted transportation routes. The challenges extended beyond urban areas, highlighting the diverse and multifaceted nature of the rescue operations.

The Emotional Toll on residents and the Nation

The seismic events, while leaving physical scars on the landscape, also etch profound emotional imprints on the hearts and minds of those directly affected and the collective consciousness of Japan.

Fear and Uncertainty:
For residents of Ishikawa, fear and uncertainty became unwelcome companions in the aftermath of the seismic upheaval. The relentless aftershocks, coupled with the looming threat of tsunamis, cast a shadow over daily life. Evacuation orders disrupted routines, and the palpable anxiety lingered in the air as communities sought refuge in government offices and evacuation centers. The emotional toll was not confined to the physical destruction but permeated the fabric of everyday existence.

Grief and Loss:

The tragic loss of lives—four reported casualties—became a somber chord in the emotional symphony that unfolded. Families faced the anguish of bidding farewell to loved ones, and communities grappled with the collective grief that accompanies such tragedies. The emotional toll extended beyond the immediate circle of those directly affected, resonating with a nation that empathized with the profound loss experienced by individuals and communities.

Solidarity and Resilience:
Amidst the emotional turbulence, "Rising Waves" captures glimpses of solidarity and resilience. Evacuees in government offices and schools found solace in collective experiences, leaning on each other for

support. The emotional fabric of the nation, too, was woven with threads of empathy and shared humanity. U.S. President Joe Biden's expression of solidarity reinforced the interconnectedness of nations in times of crisis, contributing to a broader emotional narrative of global compassion.

Chapter 4:

Analyzing the Earthquake

Insights from earthquake expert Professor Anastasios Sextos

The insights offered by earthquake expert Professor Anastasios Sextos, a guiding voice amid the seismic tumult that engulfed Japan. His expertise, drawn from the realm of earthquake engineering at the University of Bristol, became a beacon of understanding in the face of the complex geological forces at play.

Professor Sextos highlighted the peculiarity of the earthquake hitting the west coast of Japan, an occurrence less frequent than seismic events on the east coast. This geographical nuance played a crucial role in shaping the impact of the seismic forces. As the earthquake occurred offshore, the exposure of major cities to stronger shaking was significantly lower. This geographical insight added a layer of understanding to the unfolding events, explaining why the earthquake, while strong, remained less devastating than its east coast counterparts.

The distance from major cities in Japan, approximately 200 to 300 miles away, played a mitigating role in the potential impact of the seismic forces. Professor Sextos emphasized that, due to this offshore location, the earthquake's strength did not translate into widespread destruction in densely populated areas. This insight offered a glimmer of hope amid the chaos, suggesting that the consequences of the earthquakes could be managed within a reasonably small amount of time.

The seismic expert commended Japan for its leading role in earthquake preparedness. He acknowledged the country's proactive measures and its "excellent" early warning system, a crucial component in safeguarding civilian lives. The acknowledgment of Japan's preparedness became a testament to the nation's commitment to mitigating the impact of seismic events through advanced warning and response mechanisms.

Why casualties were relatively low

The intriguing question of why casualties, despite the formidable magnitude of the earthquakes, remained relatively low. Professor Anastasios Sextos, the earthquake expert from the University of Bristol, sheds light on the factors that contributed to this unexpected outcome.

The west coast location of the earthquake off the Sea of Japan played a pivotal role in reducing the potential impact on major cities. Professor Sextos explained that earthquakes hitting the west coast, unlike their more frequent counterparts on the east coast, are less common. This geographical rarity, coupled with the offshore epicenter, meant that major cities were not as exposed to the full brunt of the seismic forces. The distance of around 200 to 300 miles from densely populated areas significantly contributed to the lower intensity of the shaking experienced in urban centers.

The offshore occurrence of the earthquake resulted in reduced exposure of major cities to strong shaking. While the earthquake was robust in magnitude, the fact that it originated offshore meant that the intensity of the shaking diminished over the distance to populated regions. This geographical buffer acted as a mitigating factor, limiting the structural damage to buildings and infrastructure.

Professor Sextos commended Japan for its earthquake preparedness, describing it as a country that is "leading the way" in readiness. The implementation of an "excellent" early warning system played a crucial role in minimizing casualties. The advanced warning mechanisms allowed for prompt evacuation and heightened preparedness, ensuring that residents could seek safety before the seismic forces reached critical levels.

Analyzing the seismic activity and its distance from major cities

The seismic events centered in the Sea of Japan, off the west coast of Japan, marked a departure from the more common earthquakes along the east coast. This geographical nuance played a pivotal role in shaping the seismic narrative. The offshore epicenter introduced a level of complexity, as the seismic forces emanated from the sea, impacting the surrounding regions, including Ishikawa.

One of the key factors contributing to the relatively low casualties was the considerable distance between the seismic epicenter and major cities. The earthquakes, with their epicenter around 200 to 300 miles away, meant that the intensity of the shaking diminished as it traveled over the distance. This geographical buffer acted as a natural mitigating factor, preventing the seismic forces from reaching densely populated urban centers with full intensity.

The concept of reduced exposure to strong shaking emerged as a central theme in understanding the impact on major cities. Given the offshore origin of the seismic events, major cities were shielded from the full force of the earthquakes. This reduced exposure played a crucial role in preventing widespread devastation, allowing for a more controlled and manageable response.

The seismic analysis within "Rising Waves" integrates insights from earthquake expert Professor Anastasios Sextos. His explanations of the seismic dynamics, including the less frequent occurrence of earthquakes on the west coast and the specific geological characteristics of the region, contribute to a holistic understanding of the seismic activity.

Japan's Earthquake Preparedness and early warning systems

The country's proactive measures and comprehensive strategies have positioned it as a global leader in mitigating the impact of seismic events. This recognition underscores Japan's commitment to safeguarding its population through continuous advancements in earthquake preparedness.

A cornerstone of Japan's earthquake preparedness is its "excellent" early warning system. This system proved instrumental in providing timely alerts to residents, enabling them to take immediate actions to ensure their safety. The efficiency of the early warning system became a critical component in averting potential casualties and reducing the vulnerability of communities in the affected regions.

As the warnings flashed across television screens, residents in specific areas of the coast were advised to evacuate their homes immediately. This swift response, guided by the early warning signals, allowed individuals and communities to navigate the seismic challenges with a level of preparedness that is a testament to Japan's commitment to protecting its citizens.

The acknowledgment of Japan's earthquake preparedness extends beyond national borders. U.S. President Joe Biden, in expressing solidarity, highlighted the deep bond of friendship between the United States and Japan and underscored Japan's resilience in the face of the seismic events. This global recognition reinforces the significance of Japan's preparedness measures on an international scale.

Chapter 5:

Fires and Landslides

The risk of fires and landslides following the earthquakes

Aerial photos captured the extent of a massive fire sparked by the seismic events in Ishikawa City. The flames, towering and unrelenting, added a layer of complexity to the already challenging landscape. Firefighters found themselves contending not only with collapsed buildings but also battling to contain the inferno that threatened to consume several structures. The searing flames became a visual manifestation of the heightened risk of fires triggered by the earthquakes.

The seismic upheaval, with its power to crumble buildings and disrupt infrastructure, amplified the risks of fires.

Falling debris, structural damage, and the disruption of utilities created a volatile environment conducive to the outbreak and rapid spread of fires. "Rising Waves" captures the urgency with which emergency services responded to the dual challenges of collapsed structures and the looming threat of massive fires, creating a vivid portrayal of the intertwined dangers faced by communities.

The seismic forces unleashed landslides and road collapses, further compounding the environmental risks. National Route 249, a critical transportation artery, became impassable due to a landslide and road collapse. The disrupted transportation routes not only hindered rescue operations but also heightened the vulnerability of communities to the challenges posed by the earthquakes.

The extent of damage caused by fires, especially in Wajima City

The extent of damage becomes a somber reflection of the challenges faced by the community in the wake of seismic upheaval.

Wajima City, once a bustling urban center, witnessed a profound transformation as fires swept through its streets. The relentless flames, fueled by the seismic disruption, turned familiar landscapes into scenes of devastation. The narrative captures the visceral impact of the fires as they consumed buildings and infrastructure, leaving behind a cityscape forever altered.

The fires, acting in tandem with the seismic forces, contributed to the collapse of buildings and structures. The remnants of what was once a thriving urban environment now bore the scars of destruction. The narrative within "Rising Waves" paints a vivid picture of the structural collapse, detailing the impact on both residential and commercial spaces, creating a haunting tableau of urban scars.

The urgent battle against the inferno became a defining chapter in Wajima City's post-earthquake narrative. Firefighters, facing an unpredictable and dynamic environment, grappled with the ferocity of the flames. The description unfolds the challenges faced by emergency services as they navigated through smoke and embers, striving to contain the blaze and prevent further escalation of the disaster.

As fires raged, communities faced displacement, adding another layer of complexity to the aftermath. Evacuation orders disrupted daily life, and the resilience of the community became a poignant theme. "Rising Waves" explores the displacement of residents and the collective strength exhibited by the community as they sought refuge in safe zones, forging bonds amidst the shared experience of loss and displacement.

As the narrative unfolds the extent of damage caused by fires, especially in Wajima City, it becomes a testament to the resilience of a community grappling with the dual challenges of seismic forces and the unforgiving flames. The reconstruction of Wajima City becomes a symbol of hope amidst the ashes, illustrating the indomitable spirit that emerges in the aftermath of profound adversity.

Chapter 6:

Tsunami Threat

Initial tsunami warnings and their downgrade to advisories

This dynamic interplay between cautionary alerts and measured advisories became a pivotal element in shaping the response to the seismic events.

As the seismic tremors reverberated through the Sea of Japan off the coast of Ishikawa, the Japan Meteorological Agency issued initial tsunami warnings. These warnings, with their gravity-inducing language, urged residents along the coast to evacuate immediately. The vivid yellow warning lines on television screens became a visual representation of the potential threat, signaling a critical juncture that demanded swift action.

The warnings, indicative of a risk of 3m waves or more, created an atmosphere of urgency. The narrative captures the heightened risk perception and the subsequent evacuation efforts that unfolded in the shadow of the looming tsunami threat. Residents, guided by the warnings, sought refuge in designated evacuation centers, navigating the delicate balance between preparedness and the palpable fear of an impending disaster.

However, the seismic narrative took an unexpected turn as the Japan Meteorological Agency downgraded the initial tsunami warnings to advisories. This marked a shift in the perceived threat level, and the narrative within "Rising Waves" unfolds the nuances of this downgrade. The advisories, while indicating a potential risk of waves up to 1m, carried a different tone, allowing for a measured response from residents and emergency services.

The narrative navigates through the waves of uncertainty, exploring the psychological impact of the evolving alerts on communities. The downgrade to advisories introduces a layer of complexity as residents grapple with the dual challenges of seismic aftershocks and the fluidity of tsunami risk assessments. "Rising Waves" captures the delicate balance between preparedness and the need to adapt to the evolving information landscape.

As the seismic story unfolds, the initial tsunami warnings and their subsequent downgrade to advisories become a poignant reflection of the fluidity inherent in disaster response.

Affected areas and potential risks

The seismic epicenter, nestled in the Sea of Japan off the coast of Ishikawa, becomes the focal point of the narrative. The affected areas, including Ishikawa and nearby prefectures, come under the microscope as "Rising Waves" unfolds the geographical nuances. The sprawling coastline, once serene, now bears the scars of seismic upheaval, creating a visual landscape that underscores the extent of the disaster.

The narrative delves into the potential risks that emerge as a consequence of seismic events. Fires, as documented earlier, become a formidable adversary, consuming structures and adding complexity to the already challenging recovery efforts. The heightened risk of landslides, road collapses, and the disruption of critical transportation arteries amplifies the challenges faced by emergency services and communities.

The potential risks extend to the displacement of residents, captured through evacuation orders and the establishment of safe zones. The narrative within "Rising Waves" explores how communities grapple with the dual challenges of seismic forces and the risks associated with displacement. The establishment of evacuation zones becomes a strategic response to the fluidity of the situation, providing a lens through which readers can understand the complexities of ensuring the safety of affected populations.

Beyond the physical risks, "Rising Waves" unveils the psychological toll and uncertainty that pervade the affected areas. The ever-changing alerts, from initial tsunami warnings to advisories, contribute to a sense of unpredictability. The narrative captures the emotional landscape of residents navigating the waves of uncertainty, emphasizing the importance of resilience amidst the fluidity of disaster response.

Chapter 7:
National Response and International Support

In the wake of the seismic events, President Joe Biden expressed solidarity with Japan, emphasizing the deep bond of friendship that unites the United States and Japan. His statement, woven into the narrative, underscores the global interconnectedness in the face of natural disasters. The recognition of the shared hardship becomes a testament to the diplomatic ties and mutual support between nations.

"As close allies, the United States and Japan share a deep bond of friendship that unites our people. Our thoughts are with the Japanese people during this difficult time." - President Joe Biden

President Biden's words not only extend empathy but also acknowledge Japan's resilience in the face of the seismic events. The global recognition becomes a cornerstone in understanding the collective response to natural disasters, transcending borders and emphasizing the importance of diplomatic relationships during times of crisis.

The narrative within "Rising Waves" encapsulates President Biden's assurance of support and assistance. The United States, standing as a close ally, declares readiness to provide aid to those affected by the seismic events. This diplomatic commitment adds a layer of hope to the narrative, showcasing the collaborative efforts between nations in times of adversity.

"As close allies, the United States and Japan share a deep bond of friendship that unites our people. Our thoughts are with the Japanese people during this difficult time. The United States stands ready to provide assistance to those affected." - President Joe Biden

The collaboration between Japan and international allies for support

In the aftermath of seismic tremors that reverberated through Ishikawa, Japan, a profound testament to human unity emerged as nations from every corner of the globe rallied to offer support. "Rising Waves" intricately weaves a narrative of collaboration, turning the tragedy into a symphony of shared responsibility and global resilience.

As the initial shockwaves settled, Japan found itself facing the monumental task of rebuilding shattered communities. However, it wasn't alone in this endeavor. The international community, bound by a shared commitment to humanity, responded with unwavering support.

Diplomatic ties and alliances became the threads that wove this narrative together. Japan reached out to its allies, and the response was swift, embodying the interconnectedness of nations during times of crisis. The United Nations, serving as a conduit for global collaboration, played a pivotal role in coordinating aid efforts.

From the bustling cities of North America to the historic landscapes of Europe, the outpouring of support was palpable. Humanitarian aid, medical expertise, and essential supplies became the currency of compassion, transcending geographical boundaries to provide a lifeline for the affected communities.

Statements of support echoed globally, with political figures expressing not just solidarity but a genuine commitment to stand by Japan's side. President Joe Biden's words, "Our thoughts are with the Japanese people during this difficult time," became a beacon of assurance and a testament to the enduring bonds between nations.

Chapter 8:

Looking Forward

The potential for future earthquakes

As the seismic tremors that shook Ishikawa dissipate, the specter of uncertainty looms—what does the future hold for a region scarred by the recent earthquakes? "Rising Waves" takes a closer look at the potential for future seismic activity, exploring the delicate balance between anticipation and the unpredictable forces that shape the Earth's crust.

An earthquake expert, Professor Anastasios Sextos, emerges as a guide through the intricate landscape of seismic assessments. In the aftermath of the recent events, his insights offer a glimpse into the factors that could influence future seismic activity. The narrative navigates through the scientific intricacies, painting a picture of the unseen forces that lie beneath the surface.

The seismic landscape, often characterized by unpredictability, is examined through the lens of historical context. "Rising Waves" delves into Japan's history of earthquakes, highlighting patterns and precedents that could serve as markers for the future. It becomes a journey through time, unraveling the threads that connect past seismic events to the present, and potentially, the future.

The recent earthquakes, including the one in May 2023, become key markers in this assessment. The narrative pauses to reflect on the May earthquake, drawing parallels and distinctions that offer valuable insights into the seismic behavior of the region. It's a nuanced exploration that goes beyond statistical analysis, capturing the essence of the Earth's dynamic forces.

The potential for future earthquakes, "Rising Waves" asserts, lies at the intersection of science and preparedness. The narrative becomes a call to action, emphasizing the importance of understanding the risks, fortifying structures, and, most importantly, fostering a community-wide preparedness that transcends individual concerns.

As the story unfolds, the potential for future earthquakes becomes a critical chapter in the ongoing narrative of Ishikawa's resilience. It's not just a scientific inquiry; it's a call for communities to be vigilant, to stand united in the face of uncertainty, and to navigate the unseen currents with a collective spirit that has weathered the rising waves of seismic challenges.

Ongoing efforts in disaster preparedness and infrastructure improvements

In the wake of seismic upheaval, Ishikawa stands at a crossroads, facing not just the aftermath of recent earthquakes but also the unseen challenges that lie ahead. "Rising Waves" shifts focus to the ongoing efforts in disaster preparedness and infrastructure improvements, where communities and authorities unite to fortify against the unpredictable forces that shape their destiny.

The narrative unfolds in the aftermath of the recent seismic events, where a landscape scarred by destruction becomes a canvas for reconstruction. The lens now zooms in on the proactive measures being taken to build resilience. It's a story of communities that refuse to be defined by past challenges but instead harness the lessons learned to shape a safer tomorrow.

Central to this narrative is the resilience of the human spirit. Local communities, once shaken to their core, now emerge as architects of change. "Rising Waves" captures the grassroots efforts, where individuals and neighborhoods become the first line of defense in disaster preparedness. Evacuation drills, community awareness programs, and the establishment of local response teams become the building blocks of resilience.

Parallel to these grassroots initiatives, the narrative explores the role of technology in disaster preparedness. From advanced early warning systems to state-of-the-art infrastructure improvements, Ishikawa is embracing innovation as a shield against the unseen. The story unfolds as a journey into the future, where science and technology become allies in the quest for a more secure tomorrow.

The collaboration between local authorities and international partners becomes a driving force in this ongoing narrative. "Rising Waves" navigates through the joint efforts in infrastructure improvements, showcasing how global expertise and resources are being channeled to bolster the region's resilience. It becomes a tale of collaboration, where nations unite not just in response to crises but in a shared commitment to shaping a safer, more resilient world.

As the ongoing efforts in disaster preparedness and infrastructure improvements unfold, "Rising Waves" becomes a chronicle of hope. It's not just a story of overcoming challenges but a celebration of a community that, in the face of uncertainty, is actively shaping its destiny. The waves of resilience, once rising in response to seismic chaos, now become the force propelling Ishikawa into a future fortified against the unknown.

The Role of Global Cooperation in
Addressing Seismic Challenges

In the aftermath of seismic upheaval, Ishikawa found itself not only rebuilding physically but also forging bonds that transcend borders. "Rising Waves" now turns its gaze to the pivotal role of global cooperation in addressing seismic challenges, unraveling a narrative where nations unite in a symphony of support and shared responsibility.

The narrative unfolds as a testament to the interconnectedness of nations in the face of natural disasters. In the wake of the earthquakes, the global community responded with an outpouring of support. It was not merely an exchange of resources; it was a demonstration of a shared responsibility to help those affected by seismic chaos.

The United Nations, serving as a linchpin in international cooperation, played a central role in orchestrating support. Humanitarian aid, medical expertise, and technological know-how became the tools of collaboration, emphasizing that the challenges posed by seismic events are challenges for all humanity.

As "Rising Waves" navigates through the global response, it highlights the diplomatic ties and alliances that became the backbone of this collaborative effort. Japan, in reaching out to its international allies, not only received tangible support but also solidified the bonds that define global cooperation in times of crisis.

The narrative dives into the practicalities of collaboration, showcasing how nations pooled their resources for the greater good. It's a story of solidarity that extends beyond political rhetoric—a collaboration where actions spoke louder than words.

The ongoing efforts in disaster preparedness and infrastructure improvements become a shared endeavor, where global expertise merges with local resilience. "Rising Waves" becomes a journey through the heart of cooperation, emphasizing that addressing seismic challenges requires a collective effort that spans continents.

As the story unfolds, the role of global cooperation becomes not just a response to a crisis but a narrative thread that weaves the fabric of a more resilient world. It's a reminder that seismic challenges, like the rising waves that once shook Ishikawa, can be navigated with strength, unity, and the unwavering commitment of a global community bound by the shared goal of overcoming nature's formidable trials.

Conclusion

As the echoes of seismic tremors fade and the rebuilding process unfolds, "Rising Waves" concludes with a reflection on the events that unfolded in Ishikawa, Japan, and the broader implications that transcend the immediate aftermath. It's not just a conclusion but a call to unity, a testament to resilience, and an encouragement for ongoing awareness and preparedness.

The narrative takes a moment to look back, not merely at the destruction wrought by the earthquakes but at the indomitable spirit of communities rising from the debris. Ishikawa becomes a symbol—a symbol of strength, of collaboration, and of the enduring human spirit that refuses to be defined by catastrophe.

Reflecting on the broader implications, "Rising Waves" emphasizes the need for a global perspective on seismic challenges. It's not a localized issue; it's a shared concern that transcends borders. The collaborative efforts between nations underscore that challenges posed by nature are challenges for humanity as a whole.

Expressing solidarity with the people of Japan, the narrative becomes a heartfelt acknowledgment of the resilience demonstrated by individuals, families, and communities. It's a recognition of the pain endured, the lives lost, and the strength that emerges from the collective will to rebuild.

In the closing chapters, "Rising Waves" transforms into a beacon of encouragement.

It encourages ongoing awareness—awareness not just of the potential seismic threats but of the collective responsibility to be prepared. The narrative becomes a rallying cry for communities worldwide to heed the lessons learned from Ishikawa, to fortify against the unpredictable, and to foster a culture of resilience that extends beyond the immediate aftermath of disaster.

As the final words unfold, "Rising Waves" leaves readers with a sense of hope—a hope grounded in the belief that, together, humanity can navigate the uncertainties of the future. The book concludes not just as a chronicle of seismic events but as a testament to the enduring power of unity, preparedness, and the shared commitment to weather the rising waves of challenges that nature may present.